Quality Online Learning and Testing

The Planning, Design, Development, Delivery, Evaluation, and Maintenance of Online Learning courses, and Online Testing courses.

By

Lada Prokop, MSLS

Dedication

Dedicated to my three sons, Marko, Roman, and Pavlo, who managed the household, and kept the home fires burning, while their mother worked. Dedicated to my grandson Evan,who posed for the cover.

Contents
An Overview

Contents
The Process

Introduction

History of Instructional Technology Overview

The field of Instructional Technology dates back to the entry of the U.S. into World War 2. A situation arose, when a large number of soldiers had to be quickly deployed into the field of battle, resulting in a high fatality rate. The trainers realized, that much better training techniques were required, to save the lives of soldiers, while training them quickly, and well.

Programmed instruction fit the bill. Objectives told the soldiers what they had to learn. Clear instruction taught, and demonstrated the objectives, and tests were designed to test, exclusively, the objectives taught.

The army continued experimenting, and developed robust training programs, which continue to this day.

Additionally, experiments were being conducted at Research Universities to find out, how users learn best, and what interferes with learning.

Many books in the field of Instructional Technology were published. Books by such authors as Robert F. Mager, «Goal Analysis», «Preparing Instructional Objectives», and others, as well as by Robert Gagne «Conditions of Learning», and by John W. Best and James V. Kahn «Reseacrch in Education» became classics in the field of Instructional Technology.

Before the popularity of the computer, Instructional Technology incorporated multi-media. Text, and graphics were used. The visual aspect included pictures, cartoons, photography, and television. Photography utilized still, and video cameras. In schools, overhead projectors, and other projectors were popular.

The audio component was accommodated by cassettes, tape recorders, radio, and television. Television was also applied to long distance learning. Long distance learning was used, where long distances made the proximity of schools impractical, as was the case in the Australian outback.

In 1962, the University of Illinois developed the first generalized Computer Assisted Instruction (CAI) System, and «Computer Managed Instruction» (CMI) System, called PLATO, which, after fourteen years, they sold to a computer mainframe company Control Data Corporation (CDC). Thus, business got involved with online, computer learning, and testing. The PLATO Learning Management System (LMS) was a forernner to current LMSs. This took place in the nineteen seventies, and computer communication, which today is known as the «Internet», was exceedingly expensive.

CDC charged $ 50.00 an hour for access time on their mainframes. Hence, only big government, and big business availed themselves of computer LMS. Control Data Corporation developed curriculums, and courses for their own networks, like Basic Electronics, and many others, which they sold to General Motors, where they installed them in a hundred plants, to prisons, and to other big entities. They also developed specific programs like Nuclear Safety Testing for Detroit Edison.

Eventually, years later others built Learning Management Systems mostly for sale. But educators, also offered free usage, such as MOODLE from Western Australian University.

In 1969, the United States Department of Defense created, and funded the Advanced Research Project Agency Network (ARPRANET), so that computers could communicate with each other.

The Network Control Program (NCP) used Packet Switching. It grouped data into packets sent via a medium, which were used by simultaneous communication sessions.

In 1976, the Transmission Control Protocol/Internet Protocol (TCP/IP), the basic communication language, or protocol of the Internet came into being.

In the 1980s, fiber optic cable was invented. In the eighties many companies installed fiber optic cables, even under seas, connecting continents. While, many of the companies themselves went bankrupt, the cables connecting the world remained, for use by the Internet.

In 1989 Tim Bernard Lee developed the Hyper Text Markup Language (HTML) for communication among scientists. It enabled the sharing of computer stored information over the Internet. It was open-source. In 1990, it was called the World Wide Web.

In 1993, Marc Andreessen added image capability to text HTML. In open source tradition, Tim Bernard Lee offered HTML for free. The World Wide Web became hugely popular. Currently, HTML can embed video, sound, and programs written in other languages. Simulations, and virtual reality are used for automotive, flight, and defense training. Companies, and Universities used HTML or other programs to develop Learning Management Systems.

After, the Internet was built, and communication became affordable, online learning, training, and testing became ubiquities in the U.S. Now, most big business, Universities, Community Colleges, Schools, and Government Facilities in the U.S. use online learning, training, and testing.

To supply the need, a great many Learning Management Systems (LMSs) are available online, both free, open-source, and proprietory, with a purchase price, fee, and royalty payments. LMSs have different prices, and different features.

Free open-source LMSs were mostly developed by Universities for their own, and other students, and faculty. Moodle by Northwestern Univesity of Australia, which has become quite popular, and is constantly being upgraded, ILIAS by the Univesity of Cologne, OLAT by the Univesity of Zurich, and others.

In the U.S., the Open Ed X for the «Massive Open Online Course» (MOOC) was started by the MIT «Open Course Ware Project» in 2000. Soon, other Universities like Stanford, Caltech, Harvard, Univesity of Pennsylvania, Univesity of Texas, and Univesity of California joined. Currently, Open Ed X has 100 partner Univesities. By 2010, some MOOC courses had millions of students. By 2018, 8000 courses were developed, using Open Ed X with 50 million learners. Verified Certificates are issued for course completion.

All LMSs advertise online, and provide detailed descriptions about their features, and pricing.

Additionally, online reviews are provided by other parties, under such headings as:

«5 Top Open-Source Learning Management Systems»

«15 Top Cloud-Based Learning Management Systems»

«Best for Enterprises and Training Companies», which is an online review of 9 Learning Management Systems by PC Magazine, both open-source, and proprietory. PC Magazine rates «Instructure Canvas» as «the best educational Learning Management System on the market today.»

The Forbes article by Z. Friedman, titled «The Top 7 Websites for Free Online Education» lists, and describes the following: «Khan Academy, Ed X, Coursera, Udemy, Ted-Ed, Codeacademy, and Stanford Online.» In proprietery LMS "Blackboard", and "DL2 Brightspace" are popular in higher education.

Justin Ferriman, CEO of LearnDash, under the heading: «5 Best Rapid ELearning Software Programs» reviews the following LMSs: «Articulate, Storyline, iSpring, LearnDash, Adobe Captivate, and Lectora».

Overview

Part 1

Competency Based Learning Overview

Introduction to Competency Based Learning

Main Components of Online Learning, and Testing Overview

Behavioral Objectives Overview

Practice, and Instant Feedback Overview

Testing, and Accreditation for Courses Overview

The Leaning Management System Overview

Organization of Material Overview

Scope of Learning Courses Overview

Scope of Testing Courses Overview

Competency Based Learning Introduction

Competency based learning requires an establishment of clear, precise, behavioral objectives, teaching them, and then testing these same objectives.

Online learning, and testing is no different. Actually, online provides an excellent opportunity for competency based learning, and testing. Developing online competency based learning, and testing should start with the question:"What should be learned-taught?"

Hence, what should the learner be able to do, after the learning. Goals, and objectives will answer that question. Goals are used to organize, and structure the learning in a field. They constitute the main headings. Goals define the overall purpose of the learning. Having established clear, precise, behavioral objectives, the next question is: What content, and media provide the best medium to master these objectives? Measuring learning is done during the process itself, with practice, and instant feedback. Since, the teacher is not physically present, constant interactions are built into the learning process. Tests are developed to measure the mastery of the objectives for the purposes of assigning completion certificates, and/or grades.

In competency based testing, the tests are done to correspond with the objectives. Tests, which test extraneous material, not covered by the objectives, are not competency based. Detailed feedback about passing (the measurement of the mastery of objectives) is given to the learner, and to the provider. Information about objectives that have not been mastered is also given to the learner. The learner is then, redirected to the content/skills to be learned in the course, in order to master these objectives.

There are several main phases in the process of online learning, and testing. They are:
- Planning, and Design
- Development
- Quality Assurance, and Evaluation of the material
- Delivery
- Monitoring
- Maintenance

Components of Online Learning, and Testing Summary

Behavioral Objectives Overview

Behavioral objectives define the actual learning. What is presented, the medium in which it is presented, and the degree of mastery to be achieved.

Objectives are stated in action terms: "The learner will be able to…". This, is followed by an action the learner will be able to perform. Objectives define the condition: "Given…" This, is followed by a description of the circumstances provided to the learner.

Objectives describe the standard. How well the action must be performed. The standard is important, not only for learning, but also for practice, as well as for testing. In competency based learning, practice, and tests are developed to correspond to the objectives.

For example: "Given the capitals of 12 countries, the learner will be able to match 7 capitals to the corresponding 7 countries correctly." In the above example "Given the capitals of 12 countries" is the condition.

The action required is: "The learner will be able to match the capitals to the corresponding countries."

The standard is: "Match capitals of 7 countries correctly."

Practice, and Instant Feedback Overview

Practice, and instant feedback assist in learning, and reinforce the learning.

For example, the question may appear on the top of the page, and the feedback on the bottom of the same page, allowing the learner to repeat the question, if needed.

By providing directions, to the content related to that question, the learner is allowed to review the content, if needed, for the mastery of that question. Practice, and instant feedback, not only test the learner's memory, but provide the opportunity to repeat, and to review the material, which has not been mastered. Thus, computers accommodate both the fast, and the slow learner. The intent is mastery of the content at the learner's own speed.

Testing, and Accreditation for Courses Overview

Testing can be done frequently, say after each module, or section, or after several modules, and sections, to provide the learner with additional feedback. Since, the computer does the grading, it requires little additional effort, and is useful in reinforcing learning, as well as providing feedback.

Larger tests, such as those, that measure the learning for a complete course are developed separately from the learning course. They are delivered separately from the learning, and may be used for accreditation.

A test may be required to be taken in one sitting, as is customary with accreditation tests. Since the computer does the grading it is possible to break up a large test, and to allow the learner to take parts, or sections of the test at the learner's convenience, especially if different content is covered in different sections.

Accreditation is communicated electronically to the learner, and the provider, with detailed information of objectives mastered, and not mastered. Direction is provided to content of not mastered objectives in the learning course.

Online certificates with appropriate signatures are generated for accreditation, allowing the learner to print them out. Additionally, a hard (paper) copy of the certificate may be mailed to the learner.

Learning Management System Overview

Having the objectives, and test items selected, reviewed, and verified the Programmer, together with the Instructional Designer (ISD), and the Project Manager (PM), if costs are involved, select the Learning Management System (LMS). The LMS may be commercial, and bought or free, and open-source software. Universities have developed LMS programs for sale or share them freely, as is the case with the Perth Western Australian Company that developed Moodle, and provides it without cost. Sometimes, in house Programmers with educators have developed their own LMS. For example, the Ford Company has developed its own LMS for testing professional engineers, and their accreditation. Usually, one LMS provides a program for learning, testing, and monitoring.

Organization of Material Overview

Several major objectives make up a module. Several modules, ten or more make up a course. About three learning courses, and their corresponding testing courses equal a full-time Semester or University Quarter. However, online learning, usually, allows learners to take the courses at their own speed, regardless of Semester scheduling, after all the records are retained on the computer.

Scope of Testing Courses

About thirty test items are used to accredit one course. Sometimes SMEs require more test questions. Fifty test items are considered a long test.

Overview

Part 2

Process Overview

The Team Overview

The Design, and Development Team Overview

The Course Delivery Team Overview

The Background of Team Members: Background of the Design, and the Development Team Overview

Background of the Delivery Team Overview

The Team Process Overview

The Team Overview

It is most unusual for one person to possess all the skills, and training to develop online learning, and testing. Due, to the many skills required in providing quality online learning, and testing, companies form teams of people to encompass all the skills:

The Design, and Development Team Overview

The Design, and Development Team may consist of:

•   A Subject Matter Expert (SME) with knowledge of the content, and appropriate printed material, if available. The SME works with the team to impart the content, and reviews the course for content accuracy, signing off on the final course.

•   An Instructional Systems Designer (ISD) to write the Instructional Design Document, and to translate the contents for online learning, and testing. This, includes: The course design, the writing of objectives, and test items. The ISD reviews the course for instructional soundness, signing off on the final course.

•   A Developer to enter the course online.

•   A Graphic Artist for the look, and feel of the course pages online, and for the graphics.

•   A Videographer for course media production.

•   A Photographer to provide photographs.

•   An Animation Specialist may do animations, and simulations.

•   A Programmer for selection of the required programs, to teach the team in the use of the programs, to program difficult parts, to review, and to ensure that the course delivery is smooth, without any computer glitches. The Programmer is resposible for smooth computer delivery, and signs off on the final course.

- A Project Manager (PM) for budget, and timeline observance, and adherence. The PM writes a Project Plan, which is shared with the team members, and the funding source, usually, the upper management in a company or institution. The PM may require weekly reports from the team members, call meetings on progress, remove roadblocks, and is responsible for team cooperation, and productivity. The PM replaces a team member, who has left. On rare occasions, when the team member does not perform, the PM may require assistance from upper management to bring the performance in line, or for replacement. Often, the Instructional Systems Designer has project management training, eliminating the need for a separate PM for each course.

The Course Delivery Team Overview

- The course delivery team consists of:
  Administrators for delivery, and maintenance, Learning Facilitators, and Facility Mangement.
- The Administrators keep track of learner course completion, and test scores. These, are communicated to the learner, and to the institution. Administrators also track requests for changes, which are communicated to the affected team members. An assigned Administrator emails, and mails paper completion certificates.
- An Administrator is assigned for course maintenance
- If a facility is being used, Facility Management, Cleaning Crew, and Learning Facilitators are required.

Background of the Design, and Development Team Overview

The Project Manager (PM) has a degree in Administration, and Management.
The Subject Matter Expert (SME) is an expert in the field. The SME may be a Teacher with field expertise/accreditation, or an automotive Engineer in an Automotive Company, or a Medical Doctor, or Nursing School Professor, depending on the type of computer learning, and testing to be produced.
The Instructional Systems Designer (ISD) is an Educational Specialist. The ISD has studied Research in Education, specifying, the best delivery for optimal learning, instructional design, instructional evaluation, and instructional technology. Instructional Technology is the application of technology to learning, including computers for education.
The Programmer has studied Computer Science, and Programming. The Audio Visual Team members usually have degrees in their respective fields.

Background of the Delivery Team Overview

The Programmer has studied Computer Science, and Programming. Administrators have a degree in Administration.
Facility Managers have degrees in Facility Management.
Additionally, Learning Facilitators, who are accredited Teachers, may be assigned in a facility to assist learners with any problems, and for progress monitoring.

The Team Process Overview

The Instructional Systems Designer (ISD) meets with the Subject Matter Expert (SME) to define both the content, and the scope of the learning. Thereupon, the ISD writes the objectives, and test items, which are reviewed, and signed off by the SME to ensure validity, and content correctness. The Objectives document, and the Testing document, which are developed in a word processing program, are shared with the Development Team. The word processing program may be a commercial one like Microsoft Word, or a free one from Google Office.

The Programmer together with the ISD, and the PM select the Learning Management System (LMS). The ISD meets with the Team to determine the media to be developed. If stock photography is not used, the Photographer, who may use Adobe Photoshop, or another image editor, provides online (often jpeg) photographs to illustrate the learning of the objectives. A program called Flash, now owned by Adobe Company became very popular for images, graphics, and for interactions in learning. However, other programs, and open source programs are also available. The Video Specialist will provide video clips, if required to illustrate movement, and/or assembly to be learned. Adobe in its Creative Suites offers a video editing program, called Premier. Large companies have separate Photography, and Video Departments, providing a library of photographs, and or videotapes for all the Instructional Designers in their company to choose from, eliminating the need for a separate photographer, and or videographer for each division. Education institutions have Curriculum Centers producing multi-media for faculty and students. Alternatively, the Programmer may program animations and/or simulations.

A free animation program called Alice is offered by the Carnegie Mellon University. It is designed for non-programmers, and uses object oriented programming. It was funded by the National Science Foundation. Only, media pertinent to the learning required is used, otherwise it distracts from the learning. The idea is to keep it simple. The Graphic Artist designs the look, and feel of the computer pages, called screens. This, ensures a pleasing appearance.

The Graphic Artist determines the screen colors used for the background of the screens, for the photograph, and video positions, and frames, and for title sizes. Black is recommended for the color of the type, as it is the easiest to read, especially for people with visual problems. A light background color is selected for the same reason. The ISD writes an Instructional Design document for the learning course, and an Instructional Design document for the testing course. The ISD organizes the objectives into modules, and the modules into a learning course. The documents are shared with the Team.

Usually the Team already has a Learning Management System (LMS). It may have been developed in house by their Programmers. Mostly, a commercial, or an LMS from another learning institution is used. The LMS may be bought, or a free one like Moodle.. The Programmer manages the LMS on in house servers, and assists the Team in its usage. The Project Manager (PM) requests input from the Team, how much time they require for their tasks. The PM assigns the budget, and timeline for development. The Project Management document is shared with the Team for adherence. Weekly meetings are scheduled. The development process is ready to start. A Developer enters the modules into the LMS. A Developer enters the test items for the testing course. The Audio-Visual Team develops the media, using commercial programs like Adobe Photoshop, and/or corresponding free programs. The Team hands off the completed work to the Developer to incorporate it into the modules. The completed courses are reviewed by the SME, the ISD and the Programmer. If, corrections or changes are required, these are handed to the Development Team to be executed. Thereupon, the SME, ISD, and the Programmer sign off, and the courses are ready for testing by small groups for evaluation.

Testing will show, whether some critical changes are required. The SME will review the content changes, the ISD the learning changes, the Programmer the computer changes, and determine which changes are essential. At this point, changes are expensive, as they involve the whole Team and each change appears in several places in the course. Therefore, only essential changes are given the go ahead. After essential changes are made, and verified, the Programmer puts up the courses on the in house, or Customer's Website for delivery. The Programmer tests and verifies a smooth delivery.

Overview

Part 3

Course Quality Assurance, and Evaluation Overview

Introduction

Sign-Offs Overview

Testing of Courses Overview

Delivery of Courses Overview

Maintenance of Courses Overview

Quality Assurance, and Evaluation Introduction

Before the courses are delivered to the public, quality assurance, and evaluation is conducted.

Sign-Offs Overview

Throughout the development, sign-offs have been obtained for the Instructional Design documents, for the developed courses, and for the final evaluation. These are signatures from the SME, the ISD, the Programmer, the Evaluator if available, the Photographer, and the Head of the Audio-Visual Team for their work.
If required, revisions by the appropriate Team members are done to satisfy the sign-off.
Additionally, the Project Manager signs-off on the Design documents to ensure timeliness, and budget adherence.

The Testing of Courses Overview

The courses are tested prior to delivery by a small population of learners. Again, if essential only, revisions by the appropriate Team members are made.
The idea is to keep revisions after the course is produced to a minimum, by providing a robust course in the first place. The Team by this time is working on a different course, and the production budget is being stretched.
After, testing a final sign-off occurs to ensure a smooth delivery.

The Delivery of Courses Overview

Usually, different Teams deliver the learning course, and the testing course from the Developers.

The courses may be taken online, in a place of the learner's choosing, or in an educational institution, or in a business facility.

Assigned Administrators, or Teachers keep track of learner progress, certificates, and revision suggestions. Programmers ensure proper computer functioning, and course maintenance. Programmers immediately fix any computer glitches.

The Maintenance of Courses Overview

The SME decides, whether the revision suggestions are reflecting the correct content, and whether they have merit.

After a period of time, and depending on the budget a revision may be undertaken.

The Process

Part 1

Planning and Design

The Project Plan

The Instructional Design

The Instructional Design for the Learning Course

The Instructional Design for the Testing Course

The Project Plan

After, consulting with the team about the time the team members require for completion of their tasks, and taking the project budget into account, the Project Manager (PM) develops a Project Plan.

The Project Plan includes the names of the assigned team members, a timeline for the completion of tasks, a budget usage assignment table, and meeting dates.

The Instructional Design for the Learning Course

The Instructional Design document provides a detailed description of the learning. It is the second step in the development process. It occurs, after the SME, and the ISD have agreed, and verified the objectives, and test items; the Programmer has selected the Learning Management System, and the Media Team has identified the media to be used with the objectives.

An Instructional Design document is developed for each course. It cites the goals, main objectives, all sub-objectives, the sequence of the presentation, media, practice, instant feedback, each course section and each lesson, which is called a module.

In the design, the course is broken down into sequential modules. Each module is described, listed in sequence, and assigned a title, and number. The practice items, feedback, media are described.

The introduction, welcome, ending, and directions' wording are provided. This is important, so that all the course modules have consistency, and use the same wording for smooth delivery to the learner. If, each module contains the same wording, the learner concentrates on the learning, and finds it easier to proceed.

All the objectives, sub-objectives, and the corresponding practice items are listed, and numbered.

The selected Learning Management System is described.

Everything the Developer needs to know for successful development should be included in the design document.

The Instructional Design is shared with the Team, discussed, and agreed upon.

The Instructional Design for the Testing Course

The ISD, and the SME have agreed on the content, number, and sequence of the test items. Whether, the test items are to be presented at random, which is usually the case, or in sequence. If test items will be presented at random, it is important that, each test item be a stand-alone item.

If the learner is permitted to repeat the test, for each repetition new test items are written, so that answers are not learned, parroted, and shared between learners. Thus, customarily, three test questions are developed for each item, allowing the test to be repeated. Also, the system randomly selects a test item from the three provided, to present to the learner. Again, this prevents learners from sharing the answers between them, prior to taking the test. Furthermore, the random generation offers many versions of the test.

The test item consists of a stem (the question) and several detractors (answers)

Several kinds of multiple choice, and matching questions may be used. The most frequently used are multiple choice questions. Many Learning Management Systems do not provide options for any other kind of questions.

There are, however:

• Drag and drop questions, where the learner drags one, or several objects to the target.

• Fill-in-the-blank questions, where the learner types the correct answer, selecting it from given options.

• Hot object questions, where the learner selects one, or more objects by clicking on it, or on them.

• Hot spot questions, where the learner selects one, or more regions on the screen by clicking on it, or on them.

• True/false questions, where the learner selects the true, or false option.

The Process

Part 2

Summary of the Development Process

Course Development

Objectives

The Module

Interactions

The Course

The Testing Course Development

Multiple Choice Questions

Matching Questions

The Stem of the Question

Detractors of the Question

Development of the Test

The Learning Course Development

Objectives

To cover all the material, objectives are broken down into sub-objectives, and further sub-sub-objectives, if required, as reflected in the following numbering example:

1.   ; 1.1. ; 1.1.1. ; 1.1.1.1. ; 1.2. ; 1.2.1. ; 1.2.2.1. ; 1.2.3.1. ; 1.3. ; 1.3.1. ;
2.   ; 2.1. ; 2.2. ; 2.3. ; etcetera.

The Module

The learning course has been split into modules in the design document. All the modules in the course have the same structure for consistency, and to make them easier to navigate by the learner.

Each module has a title page. Since, no human being is present, the title page includes a welcome to the leaner.

Research in education has shown that students learn better, if told beforehand, what it is they are going to study. Therefore, the next screen provides a descriptive overview of the content to be learned, while the following screen lists the objectives, that will be covered in the module. Again, this is done in behavioral terms. For example "Upon completion of this module, you will be able to….", followed by the list of objectives. Thereupon, material to teach each objective is developed. This, includes the written content, and multi-media.

Research in education has shown that students learn better with simple graphics than with complicated ones, as with simple graphics, the detail present in complex graphics, does not distract them from the main point, to be learned.

If movement is to be learned, animation, or a video demonstrating the movement is used.

The general principle is, to present, the essentials required, in a clear and precise manner.

Following each objective, or after, several, related objectives interactions are introduced. Interactions are test items, that may include multi-media, if appropriate. Feedback is provided to the learner, after each interaction, as these interactions are designed for learning, and not for accreditation. The learner is informed, whether the learner's answer is correct. In that event, the learner is instructed to proceed to the next item. If, the answer is incorrect, the learner is given another opportunity to answer correctly. For example "This is not the answer. Try again"…A different question item, testing the same content is presented, requiring a different answer, so that the skill or knowledge is tested, and not the parroting of the same answer. Depending on the budget, up to three tries may be allowed. (More items require more time, and more expense to develop). If the learner still, cannot answer correctly, the learner is connected to the screen, which has the learning material for the objective. After, reviewing the material, the learner retakes the interaction, and proceeds. The interaction grades are not retained, so that practice is not graded. This differs from test items in the testing course, which are graded for evaluation, and accreditation. Having covered all the module objectives, the end screens for the module are presented. Again, the learner is addressed in behavioral terms. For example: "You have reached the end of this module and now you will be able to…" , followed by a list of the objectives.

An acknowledgement of the learner's effort is expressed, for example: "Congratulations you have completed the module on …"

Directions for further learning, or testing are provided. "Continue to the next module on…" .

If this is the last module, directions instruct the learner "You are now ready to take the course test on….".

An acknowledgement that the learner has completed all the modules for the course is stated: "Congratulations you have completed the course on…".

The learner, and the Administrator are provided feedback on the completion record of each module, as it occurs.

Interactions

The instruction for the objectives is interspersed with interactions, requiring some additional activity from the learner.

Interactions are test items used for self testing by the learner. As such, different feedback options are provided to interactions than those provided for the test items in the separate accreditation test. In interactions, after each item, immediate, and instant feedback tells the learner, whether the answer was correct or incorrect. In interactions, the learner is allowed to repeat the test item, and if the learner chooses, the learner is informed of the place, where the instruction for the objective tested is located. This allows the learner to repeat the learning for the objective, if needed

The test items require the learner to select a correct answer from different options, and/or match similar items correctly.

Depending on the Learning Management System (LMS) used, some or all of the following types of test items, known as learning interactions are available:

• A drag, and drop learning interaction, where the learner drags one, or several objects to the target

• A fill-in-the-blank learning interaction, where the learner types the correct answer, selecting it from given options

• A hot object learning interaction, where the learner selects one, or more objects by clicking on it, or on them

• A hot spot learning interaction, where the learner selects one or more regions on the screen by clicking on it or on them

• A multiple-choice learning interaction, where the learner selects one, or more answer from multiple choice options

• A true/false learning interaction where the learner selects the true or false option

The Course

Anywhere, from ten to about twenty modules make up a course. The course also requires an introduction, including a welcoming statement, the listing of the main objectives, an estimate of the time required, an explanation of the content, and modules in the course. The learner, and the Administrator will be provided with a list of the course modules completed.

The Testing Course Development

Multiple Choice Questions

Most online tests use a multiple choice question, with four, or up to six detractors. Thus, a multiple choice question, requiring one correct answer, may have four detractors. This is the most unambiguous test item, preferred by many.

A multiple choice question, requiring two correct answers, may have five detractors.

A multiple choice question, requiring three correct answers, may have six detractors.

Generally, requiring more than one correct answer, makes the test harder, as the learner may have known one, out of the two answers required, but still fails the question, as the system is set up to grade correct, only, if both questions are answered correctly.

Research in education has shown, that people cannot remember more than seven things at the same time. Experience has shown, that a multiple choice question requiring seven correct answers in one item, is impossible for most people to pass, including the instructors, who know the material. Hence, it should never be used.

Matching Questions

There are also matching questions, which require the learner to match correct items. These, often, include graphics to be matched to correct words, or items.

The Stem of the Question

The stem comprises the complete question to be answered. The stem must include the major information in complete sentences. If, the same wording is repeated in the detractors, that wording should have been included in the stem, to avoid repetition. The stem must be clear, concise, and unambiguous.

Detractors of the Question

Each detractor should be written, so that the answer represents a clear correct, or incorrect choice.
Often, the SME points out those other instances, which are possible. Then, the detractor must be rewritten, so that another, not ambiguous detractor is used.
The number of detractors required in a multiple choice question, depends on the number of selections needed.
For one selection four detractors are sufficient, for three selections six detractors may be used. The three selections are harder to answer, as each of the three selections must be correct, for the computer to grade the question correct. Therefore, the one selection is often preferred, as it clearly shows, what the learner has mastered.

Test Development

Usually, an experienced ISD is assigned for the development of the test, as many guidelines must be observed, in order to produce good test items.

Test items must be unambiguous, as far as is humanly possible.

The stem must contain the whole question.

The detractors must have only one, clear correct choice for the answer.

All objectives in the course should be tested.

The ISD produces several versions for one test item, usually three.

Usually, the ISD enters the test items into the LMS. Sometimes, another Developer is assigned.

The Developer enters the three versions for each item into the LMS.

In the LMS, the Developer specifies which detractor(s) are marked correct, or incorrect. Often, a checkmark denotes the correct answer, while the blanks denote the incorrect choice.

In the LMS, the Developer specifies, whether the items are to be generated at random, or in sequence.

The first version is the most common, and provides a robust test. If, for example, the test has thirty questions, and one out of three versions for each question is selected at random, many varieties of the test are possible.

The test Developer enters the number of questions the learner has to answer correctly, in order to pass the test. Usually it is 75%.

In life, and death situations, 100% is required. This was the case, with a safety test in a nuclear plant, where fatalities had occurred. The plant commissioned the test to avoid further loss of life, and to shield itself from more lawsuits.

All these specifications, should have been included in the Testing Design document.

The Process

Part 3

Quality Assurance and Evaluation

Introduction

Course Review

Validation Testing

Quality Assurance and Evaluation

Quality Assurance and Evaluation Introduction

After, the learning course has been developed, including all the modules, the objectives, learning text, graphics, any other media used, learning interactions, welcoming, and congratulatory ending statements, which comprise the transitions, the learning course is online, ready for review.

After, the test course has been developed as well, the course is online, ready for review.

Course Review

The SME reviews the final course for content, the ISD reviews the course for instructional soundness, and the Programmer reviews the course for smooth computer delivery, correcting any computer glitches.

Essential changes, corrections are made by the affected develpers.

At this point usually, sign-offs, only, by the SME, the ISD, and the Programmer are required.

After, the sign-offs the course(s) is ready for limited testing by a small population, typical of the end user.

Validation Testing

A group of about 20 learners is selected to take the courses in a facility, and being observed by an SME, and ISD, to verify that the instructions are sound, and the course can be taken, on an individualized basis. Any problems, the learners encounter, are analyzed. The SME determines, whether any content discrepancies exist, and are critical enough for changes to be made. The same is true for instructional soundness, and for computer glitches.

After essential changes, if any existed, are made, and reviewed, the courses are ready for online delivery.

The Process

Part 4

Course Delivery, Monitoring, and Maintenance

Delivery, Monitoring, and Maintenance

Online Delivery to Individuals

Facilities for Online Learning, and Testing

Test Facilities Only

Delivery and Maintenance

A Programmer transports the LMS with the courses to an in house, or customer Web site.

The same Programmer, or different Programmers are assigned maintenance of the online education. At all times, the course should be functioning smoothly, without any computer problems for the user.

Delivery to Individuals Online

"Anytime, anyplace, anywhere" is the claim being made for online delivery. The learner signs into the system, and takes the course on the learner's computer, wherever that may be.

Monitoring

The institution, or company offering the course, wants to know about the progress of the course, or courses. This includes statistics on: Numbers of learners enrolled, completed modules, completed courses, completed tests, as well as incomplete modules, courses, tests, and failed tests.

Computers, enable records to be kept on an individualized basis. Results are recorded in the LMS. They are communicated to the learner, and to the institution offering the online learning and/or testing.

Often, other statistical programs are also used, which show graphical, and chart representations of progress.

A Manager, or Teacher, or Evaluator reviews the progress.

A Facilitator, may be assigned, to communicate with the learner, and to assist the learner, if required.

While the computer emails completion certificates, which are enabled, to be printed out by the learner, some institutions, additionally, mail out a paper copy to the learner with the company, or institution logo, and signatures.

Facilities for Online Learning and Testing

Some companies, (both professional, and manufacturing), educational institutions, government, rehabilitation, and prisons prefer to offer their online courses, on site. They prepare facilities for that purpose. Facilities' preparation involves planning for a building with appropriate seating space, noise and climate control, lighting, electrical outlets, internet connections, furniture such as desks, chairs, dozens of computers, some printers, and scanners.

Computers, especially old computers will not work in very cold, or extremely hot conditions. Heating, and air conditioning is required. The same of course is true for the comfort of the learners.

In the event, that a brand, new facility is being built, it will involve the following: A Builder, a Building Inspector, an Electrician, a Systems Analyst to make the computer connections, and a Programmer to load the LMS into the computers.

A manual for the required facility, as well as, a manual for the system are provided. The System Facility Manual covers the smooth functioning, and maintenence of the facility. It includes the facility requirements, and instructions, while the System Manual teaches computer installation, describes the LMS, the record keeping, and the communication requirements, both with the learners, and with the management.

The software company sends an Analyst to help the Facilitator, or Manager review the manuals, start up the program, and to conduct the first session.

While monitoring is carried out at a central location, at facilities, on site local Monitors are assigned. These may be Plant Managers, Supervisors, or Teachers. They, usually, have subject matter expertise, so that they can assist learners, who are having difficulties.

Since, the computer does the grading, often the testing is not conducted in groups, but can be done on an individual basis.

If grouping is used, in on site test situations, the Facilitators ensure that learners do not communicate with each other.

Facilities for Testing Only

There are facilities, however, that exist exclusively for accreditation, or competitions.

When, the tests are for accreditation only, then computer assistance only is provided, and not assistance with the subject matter. In most accreditation on site test situations, the Facilitators ensure that learners do not communicate with each other.

Sometimes, the online tests are printed out, and a paper, and pencil version is provided to the learners for the test. That is done only in hardship situations, as this involves extra work, of scanning the results into the computer, and removes the impartiality., which the computer provides.

Index

Index

Index

About the Author - Lada Prokop

Education

Completed a Bachelor of Arts with Honors in German at the University of Sydney, N.S.W. Australia in 1963.
Qualified for a High School Teaching Certificate from the State of Michigan, U.S.
Received a Master of Science in Library Science from Wayne State University (WSU), Detroit in 1972.
Continued studying doctoral level Instructional Technology at the WSU College of Education, Department of Instructional Technology.

Work Experience

Worked as Librarian for WSU Libraries, Education Division.
Taught introductory Instructional Technology courses to graduate students at WSU, and to undergraduate students at Mercy College, now University of Detroit in the late 1970s.
Worked as Instructional Designer for the Mercy College of Nursing, producing a competency based Nursing Curriculum with Nursing Professors, planning to retire, in order to capture their knowledge, in paper modules, for the College. The modules were sold in the College Bookstore to Nursing students.
Employed as Education Analyst in the 1980s, at Control Data Corporation (CDC), working with Subject Matter Experts from Detroit Edison (Nuclear Plant Safety), Merck, and other companies to design/develop online testing, and/or learning courses for their employees, using the CDC computer network.
Managed/trained CDC PLATO (Computer) Facility, and PLATO Managers at 100 General Motors Plants as Project Manager.
Taught Computer courses, and ESL to GED students, as High School Teacher for Ferndale Public Schools, until official retirement with the Michigan Public School Employees Retirement System in 1996.

Worked at the Ford Motor Company Visteon Programming Division in 1999.

Employed as Instructional Systems Designer at MSX from 2000 until 2006, working with Ford Motor Company professional engineers to design/develop online testing and/or learning courses in professional engineering for qualification of all Ford Motor Company professional engineers.

Retirement

Enjoys writing/teaching/developing on the computer.

www.ingramcontent.com/pod-product-compliance
Lightning Source LLC
Chambersburg PA
CBHW060625030426
42337CB00018B/3206